SAFETY TECHNIQUES FOR
RADIOACTIVE TRACERS

SAFETY TECHNIQUES FOR RADIOACTIVE TRACERS

BY

J. C. BOURSNELL
Ph.D. (London), Ph.D. (Cambridge), A.R.C.S., F.R.I.C.

*Principal Scientific Officer, Agricultural Research Council,
Unit of Reproductive Physiology and Biochemistry and
Department of Biochemistry, University of Cambridge*

CAMBRIDGE
AT THE UNIVERSITY PRESS
1958

CAMBRIDGE UNIVERSITY PRESS
Cambridge, New York, Melbourne, Madrid, Cape Town, Singapore,
São Paulo, Delhi, Dubai, Tokyo, Mexico City

Cambridge University Press
The Edinburgh Building, Cambridge CB2 8RU, UK

Published in the United States of America by Cambridge University Press, New York

www.cambridge.org
Information on this title: www.cambridge.org/9780521155427

First published 1958
First paperback edition 2010

A catalogue record for this publication is available from the British Library

ISBN 978-0-521-04288-7 Hardback
ISBN 978-0-521-15542-7 Paperback

Contents

CONTENTS

Preface

The discovery of radioactive isotopes has given to science a method of microanalysis of extraordinary versatility, but, as with other methods of analysis, the technique demands a new discipline for its successful employment. It is the purpose of this book to collect together some of the more elementary disciplines which are intended largely as a guide to the beginner; nevertheless, a few of the ideas may be of value to those already well versed in radioactive techniques.

The use of radioactive isotopes as tracers in biological work is so widespread that the author thinks it unnecessary to make an apology for the occasional biological slant which is evident in this book.

An attempt has been made to adopt a positive approach to the subject, and to avoid as far as possible a rather forbidding set of rules. In this way it is hoped that the worker entering the field for the first time will rapidly gain confidence. The contents of this book may well act as a bridge between those who are apprehensive of the subject because of the potential dangers, and those who regard any measure of precaution as probably unnecessary.

No one at the present time can be unaware that radioactive materials are dangerous. Recent authoritative pronouncements have focused attention on that danger, even with quantities of radiation that were at one time considered to be probably below the threshold of danger. Ever since 1922, when the British X-ray and Radium Protection Committee published their first report, there has been a steady reduction in the maximum permissible quantity of radiation to which a person should be exposed in any week, on the evidence, which is still far from complete, of continued research on the subject. It is significant that the term *tolerance* dose of radiation has been replaced by the term *maximum permissible* quantity of radiation.

In collecting information over a number of years for the purpose of writing such a book as this, I have been conscious of the fact that I am indebted to a large number of people for many of the ideas which are expressed here. In particular, I would like to thank the following colleagues who have recently helped me by suggesting ways in which the book may be improved or made more comprehensive: Mr J. L. Haybittle, Radiotherapeutic Clinic, Addenbrooke's Hospital; Mr J. C. Heath, Strangeways Research Laboratory; Dr R. D. Keynes, Department of Physiology; Dr A. G. Maddock, Department of Chemistry; Dr C. L. Smith, Department of Radiotherapeutics; and Dr G. H. Smith, Department of Biochemistry.

PREFACE

An earlier version of the subject-matter of this book was submitted to the General Board of the University of Cambridge as a contribution to a report, drawn up by a committee composed of the above members and myself, on the control of the use, and disposal, of radioactive materials. I am indebted to the General Board for permission to publish independently, and to the Syndics of the Cambridge University Press for their unfailing help and encouragement.

I should like also to mention the help that I have received from the late Sir Lionel Whitby and from Dr S. G. J. Hayhoe of the Department of Medicine, University of Cambridge, in the section on Blood Counts. My sincere thanks are also due to Dr H. J. Dunster of the Health Physics Division Atomic Energy Research Establishment, Harwell for suggesting improvements in Appendix II.

Finally, I would like to express my indebtedness to my wife for checking the manuscript and advising on all the tasks involved in the preparation of a book.

ACKNOWLEDGEMENTS

The author would like to thank the following for their cooperation: the International Commission for Radiological Protection and the Editorial Board of the British Journal of Radiology for permission to reproduce (in Appendix III) part of Table C viii in the *Recommendations*; the Medical Research Council and the Controller, Her Majesty's Stationery Office, for permission to reproduce (in Appendix IX) a quotation from *The Hazards to Man of Nuclear and Allied Radiations*; the Academic Press, Inc., New York, for permission to reproduce the Table in Appendix VIII; Dr J. Beattie for drawing my attention to the screw-operated micro-pipette depicted in Fig. 5, of the origin of which I am unaware, and Messrs J. W. Towers and Co. Ltd. for a drawing from which the block for Fig. 1 was made.

J.C.B.

CAMBRIDGE
July 1957

Introduction 1

Radioactive materials must be classed among the most noxious substances, for they may constitute a health hazard even when used in 'tracer' amounts. The health hazard arises in the following ways:

(1) from external sources emitting radiations, which must be guarded against, to a greater or lesser extent, with all radioactive substances;

(2) by accidental ingestion by swallowing, or by inhalation of radioactive gases or vapours, dust or powder; and

(3) by contamination of the person (skin, clothing or hair).

Apart from the above health hazards, the weight of the isotope employed is usually so small and the measuring instruments so sensitive that a degree of cleanliness in the laboratory, beyond that normally considered necessary, is essential to avoid radioactive contamination, since if such contamination occurs it will sooner or later invalidate the results of experiments.

The health and contamination hazards are not separate concepts but are very closely interrelated; if either is neglected, both suffer. Rigorous cleanliness avoids contamination of both apparatus and person.

Health Hazard— 2
External Radiation

I. TYPES OF RADIATION AND UNITS

Radioactive substances emit radiations which interfere with essential processes within the living cell. The extent of this interference, depending upon the quality and quantity of the radiation present, is associated with the amount of ionisation or energy dissipated in the tissue cells by the passage of the radiation. This ionisation is measured in terms of a unit called the 'röntgen' (r.). (See Appendix I for a definition of this and closely related units.) The energy dissipated in the tissues is referred to as the 'dose'.

The types of radiation encountered are:

(a) *α-particles*, charged helium nuclei of great ionising power and little ability to penetrate tissue;

(b) *β-particles*, negatively or positively charged electrons of moderate ionising power and intermediate ability to penetrate;

(c) *electromagnetic radiation*, which may be arbitrarily divided into:

    (i) γ-rays of low ionising power and great penetration;

(ii) characteristic X-rays (of lower energy than in (i)). It should be realised that apart from this type of radiation, almost all radioactive materials emit photons of electromagnetic radiation (e.g. *Bremsstrahlung* (very soft X-rays) from pure β-emitters), and these can sometimes contribute to the radiation hazard.

Within each class, the ability to penetrate tissue is dependent on the radiation energy, the unit of which is called the 'electron-volt' (eV). A more convenient multiple of this ($\times 10^6$) is usually used—the mega-electron volt (MeV). Occasionally the multiple kilo-electron volt (keV) is employed (10^3 eV). The radiation is termed 'hard' if the energy is above about 1·5 MeV and 'soft' if it is below about 0·3 MeV.

The quantity of an isotope is referred to in terms of the curie (c.) or its sub-multiples, millicurie (10^{-3} c.) and microcurie (10^{-6} c.) (mc. and μc. respectively). Although at one time this unit referred exclusively to radium and radium emanation, the curie is now taken to be the amount of radioactive substance in which the number of disintegrations per second is $3·7 \times 10^{10}$.

Few isotopes decay by the simple expulsion of one type of radiation of definite energy. Their decay schemes are sometimes complicated processes, and as a result one disintegration may liberate β- and γ-rays of several different energies (see Appendix III, where

3

some of the radiations are listed). The 'dose-rate' is a function of the *rate* of energy release within the tissues caused by the passage of the ionising radiations. It is not possible to relate directly the dose rate within a tissue to the activity of the responsible source, or even to the number of particles passing through the tissue, without a detailed knowledge of the decay scheme of the isotope. For this reason particle counting devices incorporating a Geiger-Müller tube are unsuitable for the measurement of dose rates.

The α-emitters, because of their intense ionising power, are among the most noxious of the isotopes. The radio-toxicity of many is also increased by their long 'half-life' (the time required for the quantity of a radioactive isotope to decay to half its value) and a selective localisation in the body combined with a long *biological* half-life. As against this, because of their feeble penetrating power, they do not constitute an external radiation hazard. α-emitters are not frequently used in biological investigations, and for this reason they are not dealt with very fully here. Their use demands the most stringent application of the principles outlined in this manual, particularly with regard to control of dust, contamination and disposal.

2. FILM BADGES AND IONISATION CHAMBERS*

The International Commission on Radiological Protection (I.C.R.P., ref. 34), recommends that the general body radiation dose rate should be kept as low as possible at all times, and that it should not exceed 0·3 r. per working week. This is stated to be a radiation dosage rate which, in the light of present knowledge, is not expected to cause appreciable bodily injury at any time during the worker's lifetime. It is certainly possible to keep well within this value with quantities of radioactive substances normally used in biological tracer work, provided that the precautions outlined in this manual are observed.

When working with radioactive materials, even at the tracer level, it is essential to ensure that the dose of 0·3 r. is not exceeded in any one week.† Film badges or ionisation chambers, or both, must be worn continually by everyone liable to exposure when radiation is present in the laboratory. The only exceptions are when α- or soft β-emitters (such as

* Ionisation chambers, which are designed to fit in the pocket, contain an electrically insulated charged electrode. Penetrating radiation causes a discharge of the electrode which is proportional to the ionisation caused by the radiation. They are of two types, indirect or direct reading. The loss of charge by the indirect type is measured by a separate valve electrometer, whereas the amount of discharge of the direct type may be observed at any time merely by viewing the position of a fibre on a calibrated internal graticule. This direct-reading type is often referred to as a dosimeter.

† See, however, Appendix IX.

^{14}C or ^{35}S) are employed, as radiation of such low penetrating power is absorbed by the intervening glass, air, etc. Each film badge or instrument by which the total dose received is measured *must* be changed for a new one regularly, preferably every week. For continuous work with γ-radiation an ionisation dosimeter is a reliable alternative, but such an instrument is less sensitive than a film badge for β-radiation.

It is suggested that the film badge should be worn mounted on a watch strap on the inside of the wrist to determine the radiation dosage at that point. (It is also possible to wear the badge fastened to a ring on the finger if a determination is required even closer to the source.) The film badge may also be carried in a holder pinned to the laboratory coat. Unless one is instructed otherwise, this holder should be made of a material composed of elements of low atomic number, and must expose the pink paper wrapping outwards to the radiation with no intervening material. The badge must not be kept in the pocket where it is partially screened by the material of the coat, and sometimes completely screened by metal spatulas, etc.

It is absolutely essential for the person in charge of radiactive isotopes in a laboratory to keep a detailed record in a permanent book. Successive columns should record the number of the badge, the signature of the recipient, the date of issue, the date of return of the badge, the radiation dose received, and, finally, any special remarks. It is most advisable to have the signature of the recipient recorded in the book as well as on the identification card normally issued by the radiological authority who process the badges. The record book may also be used to enter the ionisation dosimeter results where these are used, and also, where necessary, results of blood counts (i.e. whether 'normal' or otherwise).

3. BLOOD COUNTS

Usually the first sign of over-exposure of the whole body to radiation is a fall in the blood leucocyte count. The recommendations of the I.C.R.P. on this point are as follows:

Provided that radiation monitoring* (both site monitoring and personnel monitoring) is carried out in all circumstances involving occupational exposure to ionising radiations (external or internal) then

(A) routine blood counts are unnecessary in the case of workers who receive doses not exceeding one-third of the permissible doses;

* A monitor indicates quantitatively or qualitatively the presence of radioactive material. These instruments are frequently portable. The quantitative varieties measure the approximate number of 'counts per minute' by the position of a needle on a dial. Many also give audible warning by clicks on a loudspeaker or earphones. The qualitative instruments merely give the audible warning.

Air-monitoring is carried out by passing a known volume of air through a filter or electrostatic precipitator. The particles thus collected are assayed for radioactive materials with a quantitative type of monitor under standard conditions.

(B) routine blood counts are optional in the case of workers who receive doses between one-third and two-thirds of the permissible doses; and

(C) routine blood counts are desirable in the case of workers who receive doses exceeding two-thirds of the permissible doses.

However, because of the difficulty of classifying workers strictly into the above categories, the following arrangements are in operation in Cambridge and elsewhere. Workers who are expected to come into category (A) would usually have annual, those in category (B) bi-annual, and those in category (C) quarterly blood counts.

In view also of the difficulty of determining whether a slight fall in blood count has occurred, it is necessary that any worker who is likely to be involved in tracer work for any length of time or to be exposed to a radiation dose more than the maximum permissible should have at least one haematological examination before he starts work. Only by this means can even a moderately satisfactory 'normal' blood count be evaluated for that person. Interpretation of the results requires very considerable experience in medicine in general and in haematology in particular.

Later in this manual various devices and methods of working are suggested for reducing the external radiation received by the various parts of the body.

Health hazard—contamination of person   3

I. ACCIDENTAL INGESTION

(*a*) *By swallowing.* A likely mode of ingestion of isotopes is by accidentally swallowing a radioactive solution when pipetting. For this reason such solutions, however dilute, should *never* under any circumstances be pipetted by mouth. There are various types of pipette in use in which suction is applied manually, and these are dealt with in Appendix VI.

In many cases known dilutions of a radioactive solution may be conveniently prepared by weighing drops of the solution and diluent from Pasteur pipettes into a weighing bottle.* Alternatively, drops of the radioactive solution may be weighed into a measuring flask, which is subsequently made up to the mark.

No smoking should be permitted in a 'hot' laboratory, as there is a considerable danger of transferring radioactivity from the bench or hands to the lips and mouth by the cigarette or pipe. Even in a 'cold' laboratory always place these on an ash tray and never directly on the bench.†

* See section on the use of these pipettes, p. 28.
† See section on 'Contamination and Cleanliness' (p. 15) for definitions of 'hot' and 'cold'.

9

For the same reason no eating or drinking should be permitted in the 'hot' laboratory. Water should only be drunk in a 'cold' laboratory from a tumbler exclusively reserved in a place where there can be no possibility of its becoming contaminated.

Another potential source of danger in this respect is the conventional mouth-operated wash bottle. The plastic bottles which merely require squeezing should be substituted.

Finally, no glass-blowing *by mouth* should take place in a radioactive laboratory. The blowpipe, bench, tools and glass tubing should be in a place which is quite free from radioactivity.

(*b*) *By inhalation of gas or vapour.* Where gases or vapours are likely to be encountered, it is absolutely essential that they should be enclosed in a completely airtight apparatus, preferably made entirely of glass. It is usually possible to carry out the reaction in such an apparatus at a pressure slightly less than atmospheric. In this way, any leak results in air entering the apparatus and the escape of radioactive gases is avoided. If any gas or vapour is likely to be produced as a by-product from a reaction normally carried out in an open system (for example, micro-organisms in culture) proper means should be devised for its effective trapping.

At this point it is important to stress the utmost necessity of running a full-scale 'dummy' experiment with inactive material before embarking on the radioactive one. The 'dummy' experiment should be conducted from beginning to end as if one were dealing with radioactive material; only by this means can difficulties be appreciated which might prejudice the success of the radioactive experiment.

As a further safety precaution, any apparatus fitted up for conducting a reaction evolving gases or vapours should be entirely assembled in an adequate fume hood or chamber to avoid the consequences of possible fracture of the apparatus. Unfortunately, the majority of fume chambers are very inefficient extractors of air. They usually produce considerable eddies and are frequently interconnected with others which are exhausted by a common fan. Such a system can lead to a spread of contamination. Their efficiency should be tested with a smoke bomb before use.

The ideal fume chamber should provide easy access to the apparatus, be free from external eddies with an adequate, but not excessive, rate of flow of air, and should discharge its effluent in a position well away from windows, etc. In this way there can be no possibility that the diluted effluent is breathed by other persons. Finally, all the services to the fume cupboard should be controllable from the outside.

(c) *By inhalation of dust or powder.* A 'manipulator' box ('dry' or 'gloved' box) should be employed when radioactive powders have to be dealt with (for instance, in opening radioactive material in powder form). Its use avoids draughts, and confines the radioactive material in a small volume. The box should be constructed so that it is capable of easy decontamination. It is also useful when such material

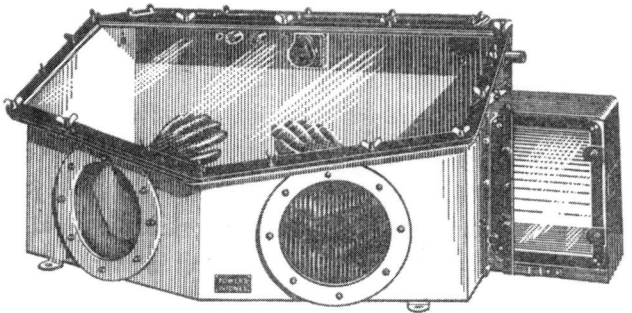

Fig. 1. A manipulator ('dry' or 'gloved') box.

as dry radioactive protein powders, $Ba^{14}CO_3$ etc., are being manipulated, since these are easily scattered by even a slight draught; it may also be used for working in a dry or inert atmosphere (in a damp atmosphere $Ba^{14}CO_3$, for instance, slowly interchanges its radiocarbon with aerial carbon dioxide). It is an advantage if the pressure inside the box is kept slightly below atmospheric, for the reason already pointed out.*

A manipulator box is essential when α-active material is being dealt with. In this case it is necessary

* It can be seen that for some purposes a manipulator box provides a convenient and economical alternative to the fume chamber.

to filter the outgoing air. Air monitors should be used to ensure that there is no possibility of inhalation of air-borne active dust.

These remarks also apply to the use of large quantities of soft β-emitters (which cannot be adequately monitored otherwise) and of millicurie amounts of isotopes of high toxicity (see Appendix III) when there is any likelihood of a dust hazard arising from their employment.

2. CONTAMINATION OF HANDS

It is sometimes difficult to avoid the adsorption of traces of radioactive material on the skin of the hands. This adsorption, which can be very persistent, may be largely prevented by the prior treatment of the hands with a suitable barrier cream, and one of these preparations should always be used where there is any possibility of skin contamination. If contamination occurs which cannot be removed by soap alone, the contaminated area should be scrubbed with a nail brush and soap. If this method fails, pumice stone should be used, and in particularly resistant cases the area of skin may in some cases be rubbed with the inactive form of the radioactive contaminant. The victim should not be satisfied until all the radioactivity has been removed, or at least until the level of activity has been reduced to below the maximum permissible value. (The figures for this are given in

Appendix II, which also sets out the corresponding figures for feet, clothing, bench top, etc.)

In this connection a ready supply of paper towels avoids the possibility of contaminating the normal laboratory towelling. The difficulty of removing contamination from the hands is considerably increased if the skin is rough, and it is wise to keep the skin soft by the use of a good-quality hand cream, especially as the hands are likely to be washed much more frequently than in a normal laboratory. The provision of a good-quality toilet soap is also an advantage in this respect. It is most advisable that the finger nails should be kept short and that radioactive isotopes should not be handled by anyone with an open wound on the hands or wrist, even though this is adequately bandaged. There is, under these circumstances, the possibility of radioactive material entering the blood stream.

When the protection of the hands from contamination by the use of thin rubber gloves becomes necessary, the worker should be familiar with the method of putting on and removing surgical gloves which avoids contaminating the hands in the process. This method is described in Appendix V. If it is necessary to touch gas, water or electricity controls with the contaminated gloves, the paper towels mentioned above may be conveniently used to prevent the direct contact of the gloves with the taps or switch.

Contamination and cleanliness

4

1. 'HOT' AND 'COLD' LABORATORIES

There seems to be no clear definition of these terms, but they may perhaps be differentiated as follows:

(i) A hot laboratory is a place where an active consignment is opened and dispensed (for subsequent use in a cold laboratory) and where, frequently 'carrier free' material* is dealt with. In this laboratory, both the health and contamination hazards may be serious. The peculiar problems in hot laboratories handling curies of radioactive material by remote control will not be considered. The maximum amount of any open source of β- or γ-emitting isotope which may normally be dealt with in safety using the precautions outlined in this manual is 50 mc. The lower activity limit for which a hot laboratory is necessary depends partly on the radiotoxicity of the material. This should not, however, be the only criterion, for it is desirable to separate high

* A radioactive isotope, in elementary form or combined, which is not mixed with the normally occurring (stable) counterpart. Where admixture occurs, the stable form is known as the 'carrier'.

15

and low specific activity material, whatever the absolute activity or toxicity, in order to minimize contamination hazards.

(ii) A cold laboratory deals usually with radio-active material diluted with carrier or in other ways. Consequently, the health hazard is considerably less than in the hot laboratory, but a contamination hazard is still present.

It is most desirable that hot and cold laboratories should occupy different rooms. If they are unavoidably situated in the same room, the portion devoted to hot work should be partitioned. This obviates the otherwise extreme danger of radioactive contamination influencing the final experimental results. The hot laboratory should have a complete and exclusive set of glassware always kept there. Carrier-free isotopes are very likely to become strongly adsorbed on glassware, and it is frequently difficult or impossible, and certainly uneconomical in time, to remove them. For this reason the glassware used here should not be returned into the general circulation of the laboratory. When several different isotopes are employed it may be desirable to use separate sets of equipment for each.

Polythene apparatus is being used more and more in radioactive work, as this material is generally less liable to contamination than glassware; moreover, it is not subject to fracture. The range of useful articles which may be purchased is steadily increasing. The only disadvantage of polythene apparatus is that, at present, it may not be heated above 70° C, although the new high-density polythenes such as 'Alkathene HD' may be used for short periods up to 110° C.

The hot laboratory should also have special apparatus which is designed to provide the worker with the maximum possible protection from radiation.

Essentially this should comprise:

(1) Trays of stainless steel or plastic (bakelite or polythene). These trays should be quite shallow and have rounded corners for easy cleaning.

(2) Rubber gloves and rubber finger stalls or 'thimbles'.

(3) Polythene sheet for spreading on the bench as a temporary working surface.

(4) Stainless steel forceps (long sponge and Spencer Wells types).

(5) 2 in. thick lead interlocking bricks.

(6) $\frac{3}{8}$ in. thick plate-glass screen (about 15 in. square) in a frame which may be clamped to the front of the bench.

(7) Lead test-tube holders.

(8) Barrier cream for the hands.

(9) A portable monitor for detecting radiation.

(10) Absorbent tissue in dispenser.

2. WORKING SURFACES

The purpose of the trays and polythene sheet is to minimise the effect of accidental spilling of active material. Such a 'spill' may be mopped up more readily from an impervious surface than from the wooden bench top. It is important that such spills should be completely cleaned *immediately* they occur, otherwise further and more serious accidents will almost certainly ensue. For this purpose, a liberal supply of good-quality absorbent cotton wool, already divided into pieces, should be kept at hand in a screw-capped wide-mouth container such as a confectioner's sweet bottle. The pieces may, if necessary, be held by the sponge forceps to avoid contamination of the hands while cleaning. The used pieces may be conveniently disposed of in the waste bin. For some purposes a supply of absorbent sheets of tissue is useful, particularly where the fibres left by cotton wool may be objectionable.

The choice of stainless steel or plastic trays is a matter of personal opinion. Cheap plastic trays may be thrown away if they become seriously contaminated. Equally the polythene sheet is a non-permanent piece of equipment. However, some authorities feel that too much non-permanent equipment tends to engender in research workers a familiarity leading to contempt. An initial light application of silicone polish or wax may be recommended for reducing the penetration of spilled material into the surface of trays, etc. Under certain circumstances a dust hazard can arise from a spilled radioactive solution drying on such a surface and being dispersed by air currents. When this dust hazard is likely to be present (particularly with α-active or large quantities of β-active substances) it may be prevented by lining the tray with absorbent paper before the experiments. This paper should be renewed frequently.

By use of these trays the possibility of contamination of laboratory benches is made very small. Nevertheless, to reduce the unpleasant effects of a spill on the surface of the bench itself, it should be kept highly polished with a silicone wax or, better still, covered with a linoleum of good (industrial) quality, again well polished. The linoleum, which should be stuck down, should be cut in such a way that the joins occur in little-used parts of the bench. Stone benches should be covered with quarry tiles; the more porous concrete tiles should be avoided for this purpose. The tiles should be pointed with an impervious cement and kept highly polished. Occasionally it may be advisable to cover them with strippable paint which may be easily removed and replaced.

Well-polished linoleum is probably the most effective means of rendering the floor impervious to radioactive solutions spilled on it. The number and position of joins should be carefully considered in relation to the most frequently used parts of the floor.

In this way the horizontal surfaces of the laboratory are covered with a non-porous material which may, if necessary, be fairly easily replaced. Walls, ceiling, woodwork, etc., should be covered with a light-coloured hard gloss paint. Other surfaces liable to contamination, particularly inside the fume chamber, should be covered, if necessary, with strippable paint.

3. CONTAMINATED APPARATUS

It is very important to deal with contaminated apparatus as soon as it is finished with. All glassware and polythene ware should be washed out at once in water and then immersed in water, containing a little of the appropriate carrier material. If this habit of *immediate* treatment is acquired, it will save a great deal of time by preventing radioactive material from drying on the glass or polythene and thus becoming much more difficult to remove subsequently. Care should be taken to ensure that radioactive material is not accumulated in the chromic acid or other cleaning fluid. Rubber bungs, tubing, etc., may be cleaned by boiling in a solution of Calgon (sodium hexametaphosphate) containing a trace of carrier.

Before being put away for further use, every article which has been contaminated should be inspected with the monitor, although it should be remembered that an external monitor may not detect contamination inside glassware by substances emitting soft radiations, and certainly will not detect the presence of α-emitters. If radioactivity is still present, the apparatus should be re-treated with carrier and re-cleaned until no more activity can be removed. The subsequent treatment depends upon the level of activity and the half-life of the material adsorbed. It is useful to have a metal bin with a lid in which apparatus stubbornly contaminated with a short-lived isotope may be stored for decay. The bin might be permanently labelled in paint: 'RADIO-ACTIVE APPARATUS—DO NOT REMOVE BEFORE DATE ON LABEL', and each article placed therein should be clearly labelled with the date of the con-tamination, the name of the isotope and the date on which the radioactivity will have decayed to a safe level.

It is useful to have a supply of small ($2\frac{1}{2} \times 1\frac{1}{2}$ in.) tie-on labels which may be quickly attached with a rubber band (if they are required for long periods string must be used). If these or self-adhesive labels are used throughout the laboratory, the danger is avoided of accidental ingestion of radioactive material occasioned by licking the normal gummed labels.

A convenient accessory is a small rubber stamp by which the word 'RADIOACTIVE' may be impressed on these labels, etc. (The labels should always be torn up when they have fulfilled their function and should never be re-used.)

4. RUBBER GLOVES AND THIMBLES

There is some difference of opinion about the use of thin surgical, as opposed to thick rubber, gloves for protecting the hands from radiation. The smaller radiation dose received by the skin of the fingers through the thick rubber may be offset by the lessened manual dexterity caused by their use, which tends to increase the time of close handling. Wearing office thimbles (made of thick rubber with pimples on the surface, and sold by most stationers), covering only the tips of the fingers and the thumb, combines many of the advantages of both types of glove.

5. LEAD TEST-TUBE HOLDERS

The lead test-tube holders (see Appendix IV), which have been used by the author for a number of years, are a simple safety device which largely prevent accidental spilling or breakage, and at the same time offer some protection to the hands while carrying the solution. These containers can easily be made by a plumber from thick-walled lead tubing. The bottom of the holder should always be lined with a pad of cotton wool, and it is sometimes advisable to pack the cotton wool tightly around the sides of the test-tube, especially when the material is required to be stored, for instance, in the refrigerator.

6. RADIATION SCREENING

The lead bricks (for γ-emitters) and the glass screen (for β-emitters) give some protection to the user, and much manipulation can be carried out from behind them. One very convenient type of interlocking lead brick is depicted in Appendix IV. The interlocking is designed to screen the operator from radiation which may penetrate the crevices in a wall of conventionally shaped bricks.

It must not be forgotten that one of the most efficient methods of ensuring protection from radiation is distance alone (the intensity of radiation decreases as the square of the distance from the source). Merely moving a sample of ^{24}Na (which emits γ-rays of 2·7 MeV) to twice the distance from the operator affords as much protection as interposing a 3 cm. thick lead shield. It is therefore very wise to keep radioactive solutions, about to be used in an experiment, at the back of the bench where, in addition, they are less likely to be knocked over. It is this principle of the protective effect of distance which forms the basis of the design of the skirted glass-stoppered container shown in Fig. 4 (Appendix IV).

It is useful to state here the limits of protection given by a 2 in. lead wall. If a general body dose rate not exceeding 0·0075 r. per hour (0·3 r. per 40-hour working week) and a comfortable working distance of 30 cm. are aimed at, then:

(1) 2 in. of lead is adequate for 50 mc. of an isotope emitting γ-rays of energy not greater than 1·0 MeV.*

(2) Doubling the wall (giving 4 in. of lead screen) makes it safe to work with energies as great as 4·0 MeV emitted from 50 mc. of an isotope.*

For some purposes closed sources of radiation are used. These often exceed 50 mc. in strength. The safe working distances and shielding required for these higher activities may be calculated from the figures given in the table in Appendix VIII.

Consideration should be given to the necessity of shielding in directions other than those directly towards the worker. For instance, people in an adjoining room may not be sufficiently protected by the intervening wall of the building. Attention should also be paid to the adequate screening of the workers' legs and feet. This may require the use of lead bricks laid horizontally under the source.

* The assumption is made, in calculating these figures, that only one γ-ray photon is emitted per disintegration. Many detailed disintegration schemes are given in references 18 and 30.

7. PROTECTION OF THE EYES

The temptation to look into the open neck of a vessel containing a radioactive substance should be resisted. Apart from the possibility of splashing, the lens is particularly susceptible to damage by radiation, which under certain circumstances is canalised by the shape of the vessel. If it is essential to do this, a mirror should be used and the illumination increased, if necessary. If for any purpose *mechanical* shielding of the eye is required, a complete shield of transparent plastic, which gives unobscured vision in every direction, is preferable to goggles. It must be remembered, however, that the radiation protection afforded by both these types may be very poor.

8. EVAPORATION OF RADIOACTIVE SOLUTIONS

Extra precautions should be taken when heating a radioactive solution. It is surprising how widespread is the contamination caused by fine invisible spray from a liquid which is merely being heated; the spread of contamination is much greater if the solution is boiled. When it is necessary to evaporate a solution, and this should be avoided if possible, the solution should be placed in a closed system fitted with a condenser. Evaporation in the open, if this is unavoidable, should be conducted at the lowest possible temperature on a water-bath fitted with ceramic rings (for easy decontamination). Alternatively, infra-red heating from above is preferred by some workers as a means of reducing the spread of contamination during evaporation. If time permits, freeze drying is probably the best method for concentrating a radioactive solution.

9. OPENING OF VESSELS CONTAINING RADIOACTIVE SUBSTANCES

Radioactive materials are sent out packed in various forms. The solutions which come in vaccine-capped bottles may most conveniently be removed, under sterile conditions if necessary, by means of an all-glass syringe fitted with a stainless steel needle. It is, of course, necessary to inject air into the vessel before it is possible to remove all the solution in this way.

When sealed glass ampoules have to be opened the time of close handling of these must be reduced to the absolute minimum. A method which has been used by the author for a number of years is as follows. A number of holders are made from 2-in. rubber bungs which have been centrally bored, each with a slightly different diameter hole, to take the different sizes of ampoules. The base of the bung is stuck with rubber adhesive to a plate of lead to give stability and some degree of radiation protection underneath, when the bung is held in the hand.

The ampoule is lifted from its container by means

26

of the forceps and placed in the hole in one of the bungs which is of such a size as to make a reasonably close fit. If there is any radioactive solution trapped in the top part of the ampoule, it may be displaced by giving the holder a sharp tap on the tray. The ampoule is then quickly pushed firmly into the hole with the fingers covered with rubber thimbles. Thus the ampoule is never held with the fingers touching the part of the glass in immediate contact with the radioactive contents.

With the thimbles still worn a scratch is made on the ampoule with a sharp glass knife. This scratch is touched with a very hot small bead of glass, or a red-hot wire, heated electrically, is wound round the ampoule at the position of the scratch. These procedures almost invariably crack the ampoule cleanly at the first attempt. A sharp tap detaches the ampoule top into a small vessel contained in a lead test-tube holder,* from which it may be consigned to the radioactive waste bin when the transfer of the contents of the ampoule is completed.

Consignments of radioactive powders, which are often enclosed in screw-capped aluminium containers, should only be opened with suitable long-handled instruments in a manipulator box. A heavy steel cup keyed to fit the base of the container is a useful accessory for holding them while opening.

---

* If the ampoule is evacuated, time should be allowed for air to leak in before the top is detached.

Tests should be carried out to show (by the results of the exposure of film badges worn on a ring) that the skin of the hands receives considerably less than the maximum permissible dose of radiation during these procedures.

A word of caution should be given against the possibility of radiochemical decomposition (chemical action promoted by the presence of the radiations) producing a pressure inside closed ampoules. This is a comparatively rare phenomenon at the values of the specific activities (mc./ml.) usually encountered, and the issuing authorities take precautions, by suitably adjusting the pH of the solution, or in other ways), against such decomposition occurring in the ampoules after sealing. Nevertheless, it is not completely unknown, and for this reason it is wise to open all sealed containers in a fume chamber. For the same reason it is not safe to seal up considerable quantities of material of high specific activity for long periods.

10. TRANSFER OF RADIOACTIVE SOLUTIONS BY PASTEUR PIPETTE

The transferring of radioactive solutions may be expediently carried out by means of drawn-out Pasteur pipettes fitted with thick-walled rubber teats. With a little practice these may be used to transfer small quantities of an aqueous radioactive solution without any danger of spilling. (These pipettes are not so easy to use when it is necessary to transfer volatile organic solvents.) The pipette should never be filled to such an extent that the liquid reaches the part that is held by the fingers, nor should the teat be over-squeezed so that when the solution is actually being transferred it is necessary to continue to compress the teat in order to prevent the liquid being sucked into the upper part of the pipette or even into the teat. One or two attempts at filling should be made to enable one to judge the compression required so that the entire contents of an ampoule, for instance, are just contained in the pipette when the finger and thumb are released from the teat.

A supply of pipettes, drawn out to a medium fineness, should always be ready for immediate use. In most cases, if the pipettes which have been used for carrier-diluted material are immediately submerged in water (without the rubber teat, which need never become contaminated) they may be freed from radioactive material at the subsequent washing. (If persistent activity remains adsorbed from carrier-free material, the pipette should be discarded.)

Spills on a well-polished and scrupulously clean surface may often be completely retrieved by the use of such a pipette drawn out very finely.

In order to avoid the danger of ingestion of radioactive material the rubber teats should never be moistened with the tongue to make them slip on to the pipette easily. It is quite sufficient for this purpose to breathe on them immediately before fitting them into place.

Apart from the ease with which radioactive solutions may be transferred by these pipettes the method has the outstanding advantage of speed. If operations of this type are carried out quickly the personal exposure is correspondingly reduced.

Disposal of radioactive waste   5

The proper disposal of the radioactive waste of the laboratory is a necessary part of the discipline of every user of radioactive isotopes.

1. SOLID WASTE

This is most conveniently collected in a separate small household refuse bin fitted with a foot-operated lid and provided with an internal enamelled steel container. This container should be lined with a stiff paper bag. (A waxed paper or polythene bag must be used if wet material is put into the bin.) The bin should be clearly marked in painted letters that it contains radioactive waste, and it should be monitored frequently to avoid the accumulation of a considerable level of activity. If this happens, or when the bin is full, the bag should be tightly tied and removed. For radioactive materials with half-lives less than 100 days the bag can be stored away, preferably in open cardboard containers, in an unfrequented part of the laboratory, such as a roof space. Each bag should be labelled with the date when the contents may be safely disposed of with the normal refuse. This date is obviously dependent on the half-life of the longest-lived isotope contained in the bag. In practice a period of about ten half-lives may be used, which allows an activity of 1 mc. (an unusually large amount for such waste material) to decay to less than $1 \mu c$.

For solid refuse containing radioactive material with a half-life greater than 100 days, all α-particle emitters, and all isotopes coming into the highly toxic category, there seems at present to be little alternative to semi-permanent storage. It is convenient if a number of laboratories combine together to maintain a locked hut, or other store, in an isolated place especially for the purpose. When this store becomes full, arrangements may have to be made, through the appropriate authorities, to dispose of the material permanently (e.g. at sea).

2. LIQUID WASTE

This should never be thrown indiscriminately down the sink. Short-lived β- or γ-active isotopes (half-life less than 100 days) in solution should be stored in bottles of suitable size until they have decayed to an activity of a few μc. only (see below and Appendix III). If, as frequently happens, the liquid contains decomposable biological material, sufficient preservative (a phenol antiseptic for instance) should be added to prevent bacterial action during the period of storage. The bottles should have well-fitting stoppers and, as with the solid waste, should carry a label giving all the necessary information about the contents, safe date of disposal, etc. They should be stored in the unfrequented part of the laboratory in a suitable well-labelled container, such as a small dustbin. They should always be transported there in a carrier so as to avoid close handling and also the possibility of accidental breakage.

It is sometimes possible to precipitate or absorb out the radioactive material from a solution. Storage of the resulting sludge is often more convenient, especially when the initial volume of the solution is large.

The radioactive solution may finally be disposed of down the sink provided that:

(i) It is not α-active.

(ii) It does not exceed in activity the values recommended in Appendix III.

(iii) Inactive carrier has been added to the solution, if possible in the same chemical form as the radioactive material. When this is not possible, the same element in inorganic form must be used. The purpose of this carrier is to avoid selective accumulation by the micro-organisms populating the drains and sewers. This is particularly important when a geologically rare element or one which is selectively accumulated in the body is being dealt with.

(iv) It is washed down with copious quantities of water.

The permission of the relevant departments of the Local Authority must always be obtained before any quantity of radioactive material is disposed of down the drains or in the garbage.

Radioactive solutions of long-lived or α-active isotopes must be treated in the same way as the long-lived solid waste.

The sink in a radioactive laboratory should be of smooth white glaze finish, free of blemishes, and should never be connected to an open system of channels and traps conveying the liquid to the main drain. Before any washing up is done in the sink, it should be carefully cleaned with a commercial cleansing powder to remove grease, and the small traces of radioactive material, released during the washing of glassware, etc., should be flushed down with adequate amounts of water. A rubber mat should never be used in the bottom of the sink as this interferes with the free flow of water down the waste pipe, and accumulates radioactivity in the sink itself and on the slime which is invariably present on these mats.

3. ANIMAL CARCASSES

In work with animals, the disposal of the radio-active carcasses poses a special problem. Depending upon the quality and quantity of the radiation, it may not be safe to dispose of them through the normal channels. Incineration of many carcasses may involve the washing and monitoring of the flue gases and the safe disposal of the ash in a manner which avoids a dust hazard. Routine burying of the carcasses which contain long-lived or especially dangerous isotopes is unsatisfactory in many ways unless suitable land can be laid aside indefinitely as a 'radioactive cemetery'.

The incineration of radioactive carcasses and other solid biological waste, where this is likely to involve considerable quantities of radioactivity over a long period of time, may only be carried out safely provided that: (i) adequate washing and monitoring of the flue gases is possible; (ii) the gaseous effluent is discharged well away from any buildings and, under certain circumstances, away from growing crops; (iii) the isotopes concerned are diluted with their inactive counterparts; and (iv) the resulting radioactive ash is prevented from becoming a dust hazard (for example, by damping with water). The safety of the whole operation is also dependent upon the proper disposal of the resulting ash.

A considerable advantage may be gained if carcasses of small laboratory animals containing short-lived radioactive isotopes (particularly ^{131}I and ^{32}P) are preserved without decomposition for a sufficiently long time to allow the radioactivity to decay to a low level, when the carcass may be incinerated in the normal way, without danger of radioactive contamination. A 'cocooning' procedure, suitable for animal carcasses not greater than 3 kg. in weight, and for portions of tissue which contain radioactive material, is described in Appendix VII and reference 4 in the Bibliography.

Laboratory administration 6
and responsibility

One technically competent person should be solely responsible in a department for advising research workers on the precautions necessary for their own safety and that of others. He should also be responsible for ordering isotopes, issuing film badges or ionisation chambers and for keeping records of these.

Every worker using radioactive isotopes is solely responsible for the cleanliness of his own room. He should never leave any radioactive material anywhere without a label. This label should show that the contents of the vessel are radioactive, what the isotope is, the worker's name or initials and the date. He should never leave any contaminated apparatus without labelling it, and he should subsequently make sure that he decontaminates it himself or that it is left in the hands of a *trained* assistant to decontaminate. By this means no radioactive contamination will ever be left for an unskilled or unsuspecting person to deal with.

It is useful to have a limited number of cards with such words as 'RADIOACTIVE—DO NOT TOUCH' printed in red on one side for indicating areas of the bench where radioactive material is being dealt with. These cards should only be displayed where there *is* radioactivity, and should always be stored or turned over immediately the radioactive substance is removed. Areas surrounding large closed sources, where the maximum permissible dose rate is exceeded, should be roped off and clearly indicated.

Summary of recommendations for safe working 7

DO NOT

1. Work without radiation and health checks.

DO

Arrange for a blood count if work is likely to be prolonged.

Wear a film badge or ionisation chamber when exposed to any but very soft radiation.

Renew or change these regularly.

---

2. Ingest radioactive material by mouth by:

(*a*) eating and drinking;

Keep all food away from a hot laboratory. Keep a tumbler in a non-active part of the cold laboratory exclusively for drinking, and *never* use a beaker for this purpose.

(*b*) smoking;

Keep, and always use, an ashtray even in the cold laboratory.

(*c*) using mouth-operated wash bottles;

Use plastic flexible wash bottles;

(*d*) glass-blowing;

Keep the blowpipe, bench and tools out of the radioactive laboratory.

(*e*) licking adhesive labels in a hot laboratory.

Use tie-on or self-adhesive labels exclusively.

---

3. Pipette radioactive solutions by mouth.

Use hand-operated pipettes (see Appendix VI).

DO NOT	DO
4. Inhale: (a) radioactive gas or vapour;	Use a totally enclosed all-glass reaction apparatus in the fume chamber with adequate ventilation. Trap all gases or vapours.
(b) radioactive dust.	Use a manipulator (or dry) box (use, if possible, with a pressure inside slightly less than atmospheric). Filter extracted air.
5. Let the hands become contaminated.	Use barrier cream, gloves or rubber thimbles.
6. Work with unscreened large sources.	Use suitable screening (lead interlocking bricks for α- and plate glass for β-emitters). Always use to the fullest possible extent the protection afforded by distance.
7. Handle vessels containing large sources.	Use forceps, tongs, protective holders, rubber thimbles, etc.
8. Throw indiscriminate quantities of radioactive materials down the sink.	Conform to the maximum permissible amounts listed in Appendix III.
9. Throw any *carrier-free* isotope down the sink.	Add carrier, if possible in the same chemical form, to the discarded isotope before placing in the sink. Wash down well with water.
10. Place radioactive waste with the ordinary laboratory waste.	Use a special container with a foot-operated lid.

DO NOT	DO
11. Contaminate the bench.	Use trays or polythene sheet.
12. Place radioactive material where it is in danger of being spilled.	Use lead test-tube holders, skirted containers (see Appendix IV) and place at the back of the bench when about to be used, and in an even safer situation when not.
13. Contaminate the ordinary laboratory towels or service taps.	Use paper towelling if there is any possibility of the hands being contaminated.
14. Leave contaminated apparatus untreated.	Immerse contaminated apparatus *immediately* in water containing carrier, and clean as soon as possible.
15. Risk contamination of apparatus.	Monitor *all* apparatus after cleaning *before* it is put away.
16. Risk cross-contamination.	Dispense all radioactive materials in a separate part of the laboratory which has an exclusive set of equipment.
17. Leave radioactive material in a situation where it is likely to endanger people.	Label every sample 'RADIO-ACTIVE' and discard the label immediately its function is fulfilled. Monitor the surroundings of sources and indicate dangerous areas.

DEFINITIONS

Curie (c.). That amount of radioactive substance in which the number of disintegrations per second is 3.7×10^{10}. Subdivisions: millicurie (mc.), 3.7×10^7 disintegrations/second; microcurie (μc.), 3.7×10^4 disintegrations/second.

Electron volt (eV). The energy acquired by an electron falling through a potential difference of one international volt. Multiples: kilo-electron volt (keV) $= 10^3$ eV; million or mega-electron volt (MeV) $= 10^6$ eV.

Half-life (τ). The time during which an aggregation of radioactive atoms decays to half its number.

Röntgen (r.). The quantity of X- or γ-radiation such that the associated corpuscular emission per 0.001293 g. of air produces, in air, ions carrying one electrostatic unit of quantity of electricity of either sign. Sub-division: milliröntgen (mr.), 10^{-3} r. The röntgen corresponds to the absorption of 83.8 ergs/g. air.

Rep. If the energy lost by ionisation *in the tissues* is the same as the energy loss for 1 r. of γ-radiation absorbed in air, the dose is spoken of as one 'röntgen equivalent physical' (rep).

Rad. A new unit, adopted by the International Commission on Radiological Units, of the amount of energy imparted to matter by ionising particles per unit mass of irradiated material. It is equivalent to 100 ergs/g.

MAXIMUM PERMISSIBLE LEVELS FOR CONTAMINATION OF HANDS, FEET AND BENCH

It is difficult to state exact values for the maximum permissible amounts of radioactive isotopes which should be allowed to contaminate the hands, feet and bench top. This contamination is usually the result of accidental release of radioactive material, but may sometimes arise from normal work. Although, perhaps, the exact maximum may be stated for any particular isotope in terms of microcuries likely to produce a certain dose rate at a specific place, a determination of the value of this dose rate is obviously out of the question in many cases. All that can be done is to measure approximately the number of counts per minute recorded by a sensitive G.M. counter (or 'α-probe' in the case of α-emitters) attached to a portable monitor. Using a thin window (mica) counter for β-particles or a sensitive γ-counter, 90 counts per minute would seem to be within the maximum permissible for large areas (e.g. 30 sq.cm.) of skin on the hands and feet. For α-emitters, the figure is certainly lower than this, even if a scintillation α-probe is used to detect the contamination.

Under circumstances where the area of persistent skin contamination is quite small (a few mm. in diameter) and only rarely encountered, a relaxation of the above figures, by a factor of not more than 10 times, could be accepted, if absolutely necessary, without serious risk.

The bench top may most conveniently be tested for loosely attached radioactive material by the now conventional 'smear-test'. This consists of hard wiping of 12 sq.in. of the bench with 2 sq.in. of filter paper (some authorities suggest that the paper should be 'lightly oiled'). The activity of the paper should not exceed 300 c.p.m. for β- and γ-radiation. It must be pointed out, however, that if loosely held radioactive material can be removed in this way, a greater spread of contamination is certain to result if it is left without further treatment. Every effort should therefore be made to remove the contamination so that the activity of the filter paper is finally zero. If persistent firmly attached activity is still remaining on the bench, that part on which it is localised should be clearly indicated and should not be used during the decay of the isotope. If the isotope is long-lived, its permanent presence is intolerable, and the appropriate section of the linoleum, tiles, etc., should be removed.

CLASSIFICATION OF RADIOACTIVE ISOTOPES BY TOXICITY

MAXIMUM AMOUNTS WHICH MAY BE DISPOSED OF DOWN THE SINK

MAXIMUM AMOUNTS PERMITTED IN THE AIR

The figures in the tables which follow are taken from the following sources or observations:

(1) The classification into 'Very high', 'High', 'Moderate' and 'Low' toxicities follows that of Dunster (1954).

(2) The values for the half-lives and the principal energies of the radiations are taken from *Nuclear Data* (1950) or Hollander, Perlman and Seaborg (1953).*

(3) The maximum permissible concentrations are taken from *Recommendations of the International Commission on Radiological Protection* (1954).

As will be seen, this list is not a complete one, and does not include a considerable number of isotopes which are now available. When these figures become available it will be possible to make these tables more representative.

(4) The volume of 5×10^7 ml. used in the calculation of the maximum permissible air-borne activity is taken to be the capacity of a room of moderate size (slightly smaller than 18 ft. × 11 ft. × 9 ft.). This maximum assumes uniform distribution, which is most unlikely to occur.

(5) The maximum permissible amounts in 88 gallons of

---

* An attempt has been made to give the energy of the most abundant radiation and the percentage of the most energetic radiation present in each case.

water (about 4×10^5 ml.) are calculated directly from the corresponding concentrations in the Recommendations of the I.C.R.P. The value of 88 gallons represents the average amount of water used per day per research worker. This figure was near the lowest average obtained from a survey of the water used during the last three years in three representative Cambridge laboratories of varied function and size.

Although the figures calculated in the tables are the maximum permissible, it is strongly suggested that every worker should try to conform to the following, even lower, limits of radioactive material disposed of in this way:

(i)	Low toxicity group	100 μc./day
(ii)	Moderate toxicity group	10 μc./day
(iii)	High toxicity group (except natural Th or natural U)	1 μc./day
(iv)	Very high toxicity group	10^{-3} μc./day

(It would be reasonable to divide larger quantities than these into smaller units, which may be disposed of on successive days.)

If these figures are observed, the hazard involved for other people is very small and the recommendations of the International Commission that 'in the case of the prolonged exposure of a large population, the maximum permissible levels should be reduced by a factor of ten below those accepted for occupational exposures' are certainly complied with. Nevertheless, it should still be the aim of every worker to decrease the dangers of radiation hazard to the very minimum. It is much better to store radioactive material to decay where this is possible than to throw it down the sink, even though the amount is below the permitted level.

46

In the disposal of radioactive material down the sink, a number of further factors have to be considered:

(i) These limits should not be exceeded except under very special circumstances. The sanction of the person in charge of work with radioactive materials in the department should be obtained before this is allowed to take place. Any worker, before he exceeds the limits set out above, must make sure that no one in the department is disposing of any isotope in the same way. The maximum permissible concentrations given in the tables must never, under any circumstances, be exceeded for the entire daily output of water from the department, which should be known.

(ii) As already mentioned in this manual, it is essential to add carrier to the active isotope *before* flushing down the sink. It is not safe to assume that adequate protection will occur if the radioactive isotope is flushed down *followed* by the carrier. In many cases the interchange of carrier-free material, absorbed on micro-organisms, with the corresponding inactive carrier is a very slow process.

For those long-lived isotopes in the very high toxicity class for which there is no stable (inactive) isotope to act as a carrier, there seems to be no ready solution to the problem of disposal except indefinite storage. This does not, of course, finally overcome the difficulty, and eventually arrangements must be made for the accumulated material to be dumped at sea, but many factors are concerned in this method of disposal, and the whole question lies outside the scope of this manual.

The radiation characteristics included in these tables are only intended as an indication of the principal radiations emitted. For more detailed information references 18 or 30 should be consulted.

A. *Very high toxicity*

Isotope	Half-life	Energy of radiation (MeV)			Maximum permissible concentration		Maximum permissible amount	
		α	β	γ	In water (μc./ml.) ×10⁷	In air (μc./ml.) ×10¹¹	In 88 gal. of water (4×10⁵ ml.) μc.	In 5×10⁷ ml. of air μc. ×10²
{⁹⁰Sr	19·9 yr.	—	0·6	none }	8	20	0·3	1·0
⁹⁰Y	65 hr.	—	2·2	none }				
{²¹⁰Pb	22 yr.	—	0·02	soft }	20	8	0·8	0·4
²¹⁰Bi	5 days	—*	1·2	none }				
²¹⁰Po	140 days	5·3	—	0·8	30	50	1·2	2·5
²¹¹At	7·5 hr.	5·9 (40%)	—	o.e.c.†	30	50	1·2	2·5
²²⁶Ra	1620 yr.	4·8	—	0·18	0·4	0·8	0·01	0·04
²²⁷Ac	21·7 yr.	4·9 (1·2%)	0·04	soft	30	0·4	1·2	0·02
²³¹Pa	3·4 × 10⁴ yr.	5·0	—	0·3	30	0·2	1·2	0·01
²³³U	1·63 × 10⁵ yr.	4·8	—	0·3	30	3	1·2	0·15
²³⁹Pu	2·44 × 10⁴ yr.	5·2	—	soft	30	0·2	1·2	0·01
²⁴¹Am	470 yr.	5·5	—	soft	30	4	1·2	0·2
²⁴²Cm	162 days	6·1	—	soft	20	20	0·8	1·0

* Also very weakly α-active.
† Decays by orbital electron capture o.e.c., resulting in the emission of a characteristic X-ray.

B. *High toxicity*

Isotope	Half-life	Energy of radiation (MeV)		Maximum permissible concentration		Maximum permissible amount	
		β	γ	In water (μc./ml.) ×10⁸	In air (μc./ml.) ×10⁹	In 88 gal. of water (4×10⁶ ml.) μc.	In 5×10⁷ ml. of air μc.
⁴⁵Ca	152 days	0·26	none	10	8	40	0·4
⁵⁹Fe	46 days	0·46 (50 %), 0·26 (50 %)	1·3	10	20	40	1·0
⁸⁹Sr	53 days	1·5	none	7	20	28	1·0
⁹¹Y	61 days	1·55	1·2 (<0·1 %)	30	9	120	0·45
{¹⁰⁶Ru	1 yr.	0·04	none	10	20	40	1·0
¹⁰⁶Rh	30 sec.	3·5 (68 %)	0·7 (17 %), 1·25 (1 %)}				
¹³¹I	8 days	0·6 (87 %)	0·36 (80 %), 0·7 (3 %)	6	6	24	0·3
{¹⁴⁰Ba	12·8 days	1·0 (60 %)	0·5	30	20	120	1·0
¹⁴⁰La	40 hr.	1·3 (70 %), 2·2 (10 %)	1·7 (77 %), 2·3 (6 %)}				
{¹⁴⁴Ce	275 days	0·2 (30 %), 0·3 (70 %)	soft	10	2	40	0·1
¹⁴⁴Pr	18 min.	3·0	2·2 (weak)}				
¹⁵¹Sm	73 yr.	0·08	0·02	800	3	3200	0·15
¹⁵⁴Eu	16 yr.	0·3 (50 %), 1·9 (10 %), 0·7 (40 %)	c. 1·2	40	2	160	0·1
¹⁷⁰Tm	129 days	0·97 (76 %), 0·88 (24 %)	0·08 (3 %)	50	10	200	0·5
{²³⁴Th	24 days	0·2 (80 %)	0·09 (20 %)	20	10	80	0·5
²³⁴Pa	1·2 min.	2·3 (80 %)	0·8}				
Natural Th	long	—	—	0·05	0·03	0·2	0·0015
Natural U	long	—	—	0·2	0·03	0·8	0·0015

C. *Moderate toxicity*

Isotope	Half-life	Energy of radiation (MeV) β	Energy of radiation (MeV) γ	Maximum permissible concentration In water (μc./ml.) $\times 10^4$	Maximum permissible concentration In air (μc./ml.) $\times 10^7$	Maximum permissible amount In 88 gal. of water (4×10^5 ml.) μc.	Maximum permissible amount In 5×10^7 ml. of air μc.
^{24}Na	15·0 hr.	1·39	2·7	80	10	3200	50
^{32}P	14·3 days	1·7	none	2	1	80	5
^{35}S	88 days	0·17	none	50	10	2000	50
^{36}Cl	4·4 × 10⁵ yr.	0·7	none	40	6	1600	30
^{42}K	12·5 hr.	3·6 (75 %)	1·5 (17 %)	30	6	1200	30
^{46}Sc	85 days	0·36 (98 %), 1·5 (2 %)	1·1, 0·9	4	0·5	160	2·5
^{47}Sc	3·4 days	0·61	—	9	2	360	10
^{48}Sc	1·8 days	0·6	1·3	4	0·7	160	3·5
^{48}V	16 days	0·7 (58 %)	2·3, 1·3, 1·0, o.e.c.	3	0·5	120	2·5
^{56}Mn	2·6 hr.	2·8 (50 %)	2·1 (13 %), 0·8 (67 %)	30	5	1200	25
^{55}Fe	2·9 yr.	none	o.e.c.	50	7	2000	35
^{59}Ni	8 × 10⁴ yr.	none	o.e.c.	40	7	1600	35
^{60}Co	5·3 yr.	0·3	1·33, 1·17	4	0·8	160	4
^{64}Cu	12·8 hr.	0·66 (19 %), 0·57 (39 %)	1·34 (0·5 %), o.e.c.	50	9	2000	45
^{65}Zn	250 days	0·32	1·1 (45 %), o.e.c.	20	4	800	20
^{72}Ga	14 hr.	3·15 (9 %), 0·9 (32 %), 0·6 (40 %), 1·5 (11 %)	2·5 (26 %), 2·2 (33 %), 0·84 (100 %)	5	1	200	5
^{76}As	26 hr.	3·15 (50 %), 3·04 (60 %)	1·2, 0·6, 2·2 (weak)	2	0·4	80	2
^{89}Rb	19·5 days	1·8 (80 %), 0·7 (20 %)	1·1 (20 %)	30	4	1200	20
^{95}Zr	65 days	1·0 (2 %), 0·37 (99 %)	0·7	6	0·8	240	4
^{95}Nb	35 days	0·16	0·75	20	2	800	10
^{99}Mo	68 hr.	1·23 (80 %), 0·45 (20 %)	0·74, 0·78, 0·85	30	5	1200	25

Isotope	Half-life						
⁹⁵Tc	4·3 days	—	0·8, o.e.c.	10	2	400	10
{¹⁰³Pd	17 days	—	o.e.c.}	50	8	2000	40
{¹⁰³ᵐRh	57 min.	—	I.T.	—	—	—	—
¹⁰⁵Ag	37 hr.	0·6	0·32 (5 %)	10	2	400	10
{¹⁰⁹Cd	40 days	—	0·4, o.e.c.}	4	0·7	160	3·5
{¹⁰⁹ᵐAg	470 days	—	o.e.c., I.T.}	700	0·7	28000	3·5
¹¹¹Ag	7·6 days	1·0 (91 %)	0·34 (9 %)	5	0·8	200	4
¹¹³Sn	112 days	1·2	0·4, o.e.c.	20	3	800	15
¹²⁷ᵐTe	115 days	0·7	I.T.	7	1	280	5
¹²⁹ᵐTe	33·5 days	1·8	0·3, 0·8, I.T.	2	0·4	80	2
{¹³⁷Cs	33 yr.	0·52 (92 %)	0·66, I.T.}	20	2	800	10
{¹³⁷ᵐBa	2·5 min.	1·2 (8 %)	—	—	—	—	—
¹⁴⁰La	40 hr.	1·3 (70 %), 2·3 (10 %)	1·65 (56 %), 2·6 (3 %)	3	0·5	120	2·5
¹⁴³Pr	13·8 days	0·9	none	5	0·9	200	4·5
¹⁴⁷Pm	2·6 yr.	0·2	none	20	0·4	800	2
¹⁶⁶Ho	27 hr.	1·8 (89 %), 0·6 (11 %)	1·4 (11 %)	5	0·8	200	4
¹⁷⁷Lu	7 days	0·5 (65 %), 0·4 (17 %)	0·2, 0·1	10	2	400	10
¹⁸¹W	140 days	—	0·8, o.e.c.	7	1	280	5
¹⁸²Ta	111 days	0·5	1·2	5	0·2	200	1
¹⁸³Re	155 days	—	0·25, o.e.c.	20	4	800	20
¹⁹⁰Ir	12·6 days	—	0·2, 0·6, o.e.c.	30	6	1200	30
¹⁹¹Pt	3 days	—	0·6, 1·5, o.e.c.	7	1	280	5
¹⁹³ᵐPt	4·3 days	—	c. 1·5, I.T.	9	—	360	10
¹⁹⁶Au	5·6 days	0·27 (5 %)	0·36 (95 %), 0·43 (5 %)	20	2	800	10
¹⁹⁸Au	2·7 days	0·96	0·41	6	1	240	5
¹⁹⁹Au	3·2 days	0·3	0·2	20	3	800	15
²⁰⁰Tl	27 hr.	—	0·4, 1·6, o.e.c.	10	2	400	10
²⁰²Tl	13 days	—	0·4, o.e.c.	50	9	2000	45
²⁰⁴Tl	3·0 yr.	0·8	none, o.e.c.	10	2	400	10
²⁰⁸Pb	52 hr.	—	0·2, 0·4, o.e.c.	20	4	800	20

I.T. = Isomeric transition or nuclear isomerism. The decay of these isotopes proceeds with two different half-lives simultaneously. γ- or X-rays or both are emitted in this process.

D. Low toxicity

| Isotope | Half-life | Energy of radiation (MeV) | | Maximum permissible concentration | | Maximum permissible amount | |
		β	γ	In water (μc./ml.) $\times 10^3$	In air (μc./ml.) $\times 10^6$	In 88 gal. of water (4×10^5 ml.) μc.	In 5×10^7 ml. of air μc.
^{3}H	12·5 yr.	0·018	none	200	1	80,000	500
^{7}Be	53 days	—	0·48 (11%), o.e.c.	20	0·3	8,000	150
^{14}C	5,568 yr.	0·16	none	3*	1*	1,200*	500*
^{18}F	112 min.	0·65	0·3 (3%), o.e.c.	200	3	80,000	1,500
^{51}Cr	28 days	—	o.e.c.	20	0·4	8,000	200
^{71}Ge	11·4 days	—	0·2, o.e.c.	20	0·3	8,000	150
^{204}Tl	72 hr.	—		9	0·2	3,600	100

* As $^{14}CO_2$.

ILLUSTRATIONS OF PROTECTIVE EQUIPMENT

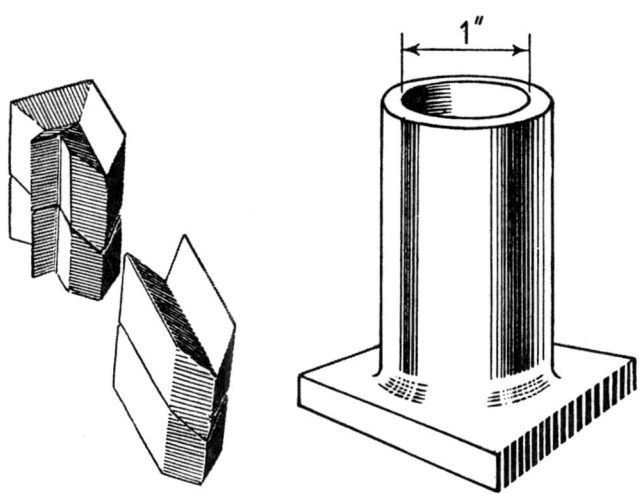

Fig. 2. Lead interlocking bricks (4 in. × 4 in. × 2 in.).

Fig. 3. Lead test-tube holder.

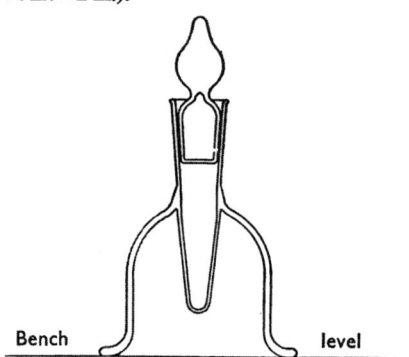

Bench   level

Fig. 4. Skirted glass-stoppered container.

METHOD OF PUTTING ON AND TAKING OFF RUBBER GLOVES

The technique employed in this procedure is such that the inside of the glove is not touched by the outside, nor is any part of the outside allowed to come into contact with the bare skin.

The procedure is as follows:

(1) The gloves should be dusted internally with talcum powder.

(2) The cuff of each glove should be folded over, outwards, for 1½–2 in.

(3) Put one glove on by grasping only the internal folded-back part with the other hand.

(4) Put the second glove on by holding it with the fingers of the gloved hand tucked in the fold and only touching the outside of the glove.

(5) Unfold the gloves by manipulating the fingers inside the fold.

(6) In taking off the gloves, seize the fingers of one glove by the other gloved hand and pull free.

(7) Take off the other glove by manipulating the fingers of the free hand under the cuff of the glove and fold it back so that an internal part is exposed which may be seized and the remaining hand freed.

HAND-OPERATED PIPETTES

1. *'Exelo' safety pipette* (0·005–100 ml.). Liquid drawn up by a type of hand-operated syringe barrel. Control of level by finger after filling. (W. G. Flaig and Sons, Ltd.)

2. *'E-MIL' auto-zero micro-pipette* (0·025–0·25 ml.). (H. J. Elliott Ltd.)

3. *'Microid' accurate pipette filler.* Rubber bulb with finger-operated valves to release liquid to mark slowly. (Griffin and George, Ltd.)

4. *Teat-operated pipette.* A side arm is attached to an ordinary pipette. To this is fitted a rubber bulb for sucking up the liquid, the top orifice being closed by the finger. Subsequent control by the finger as usual. (A. Gallenkamp and Co. Ltd.)

5. *'Pumpett' Automatic Pipette Control* (Shandon Scientific Co. Ltd.). Operation of any size of pipette by large and small rubber bulbs with micro-control screw for fine adjustment.

6. *Screw-operated micro-pipette.* A diagram of this device is shown (Fig. 5). It can be made easily and is very successful for small quantities (up to about 0·25 ml.).

7. *Syringe-operated micro-pipette* (Fig. 6). The apparatus possesses the following features:

(*a*) The tube to be filled may be slid vertically up the tube guide into position to receive the contents of the pipette, without the latter touching the sides of the tube. In this way radioactive contamination of the upper part of the tube is avoided.

(*b*) Although the apparatus is of particular value when a number of equal volumes have to be dispensed, the pipette may be rapidly changed for one of another size. A series of pipettes, capable of delivering accurately volumes between 0·05 and 0·5 ml., may be made from thick-walled tubing of different bores.

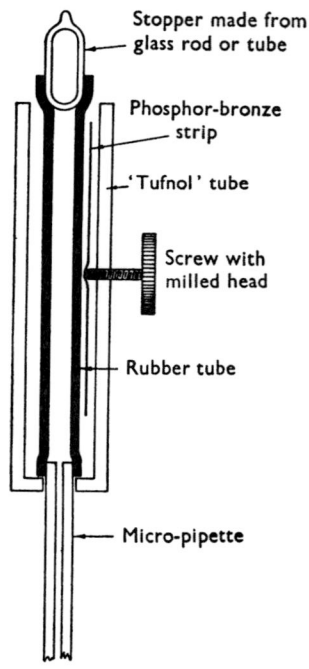

Fig. 5. Screw-operated micro-pipette.

(c) The guide may be pushed to one side if necessary when the pipette is being refilled, and then slid back again to occupy identically the same position as before.

(d) The position of the guide can be altered to suit tubes of different sizes by loosening the wing nut.

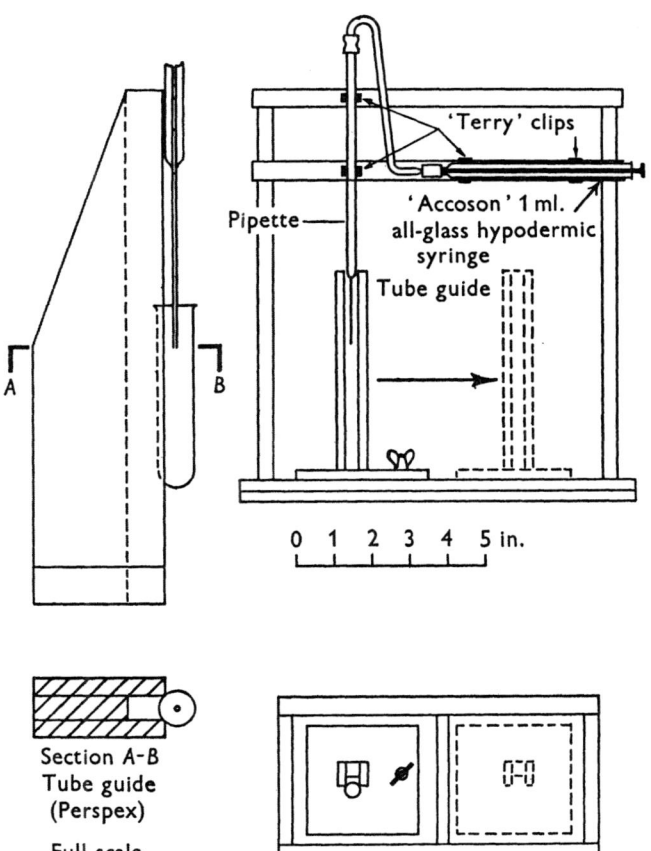

'Terry' clips

'Accoson' 1 ml.
all-glass hypodermic
syringe

Pipette

Tube guide

0 1 2 3 4 5 in.

Section A-B
Tube guide
(Perspex)

Full scale

Fig. 6. Syringe-operated micro-pipette.

57

TEMPORARY PRESERVATION OF RADIOACTIVE ANIMAL CARCASSES

Fresh bleaching powder ($CaOCl_2$) and vermiculite (a commercial expanded mica) are placed in a polythene bag and the carcass or tissue introduced. If the polythene bag is heat-sealed and the carcass tumbled about in the mixture until it is thoroughly dusted, decomposition will be prevented for a considerable time (reference 4).

Certain procedures are essential to the success of this process:

(1) It is desirable to slit the carcass ventrally along the mid-line, if this has not already been done, before placing it in the bag.

(2) If the carcass is soiled with tissue fluid or blood, it should be dusted with bleaching powder and allowed to cool before being placed in the bag. This avoids the development inside the bag of heat due to the rapid interaction of the bleaching powder and organic matter. The heating may also be avoided by applying vermiculite to the opened carcass. The finer grades of this material absorb about 4 times their own weight of fluid, and thus the interaction with the bleaching powder is slowed down.

(3) The open end of the bag (which should be made of the heavier gauges, 250–500, of polythene) should have previously been folded back in a double cuff fold outwards to protect the surfaces to be sealed from dust and organic matter.

(4) The total weight of the bleaching powder and

vermiculite should each be not less than one-fifth of the weight of the carcass. This represents, in practice, a volume of about 300 and 1500 ml. respectively per kg. weight of carcass.

(5) The animal's claws should be cut before the start of the radioactive experiment. The feet of the carcass should be bound with adhesive tape, lint or bandage before it is placed in the bag, to avoid the possible penetration of the bag. For the same reason it is unwise to sever the feet with bone forceps lest sharp splinters are left.

(6) The bag should be sealed with a suitable type of electrically heated 'soldering gun' so that only a small amount of air is included, but care should be taken in expelling the air not to blow dust on the surfaces to be sealed.

(7) The 'cocooned' carcasses should be stored in open containers to allow free circulation of air.

TEMPORARY SHIELDING FOR RADIOACTIVE SOURCES

(Reproduced by permission from H.M.Parker (1951), *Health Physics Instrumentation and Radioactive Protection. Advances in Biological and Medical Physics,* Vol. 1, p. 223. New York, Academic Press, Inc.)

This table enables one to calculate, by the addition of three quantities, with due regard to sign, and multiplication by one factor, the shield thickness for all normal values of source strength, γ-ray energy, distance of source, daily exposure time and customary shield material.

Select the column for the energy required, using the next higher energy if the exact value is not given. The entry gives the thickness, in cm. Pb, required for different source strengths at 1 m. for 8 hr./day to give 100 mr. in this working time.

Add algebraically the correction terms for working ranges or times of exposure per day, and multiply by the factor given for the shield material. Example: 500 mc. of a source emitting 1·8 MeV γ-rays at 50 cm. distance requires $(7·21 + 2·77 - 1·39) \times 1·43 = 12·3$ cm. Fe for the radiation dose to be reduced to 100 mr. in 4 hr.

The table is computed on the assumption that each disintegration yields one gamma photon of the selected energy. This leads to inaccuracies whenever the disintegration scheme is complex. More accurate calculations can be made when the disintegration scheme is known.* The increased effective transmission of shields with wide

* Many detailed disintegration schemes are given in references 18 and 30.

beam irradiation is also ignored. The table is a useful guide for the erection of temporary shielding.

It must be noted that the value of 100 mr. per day, which was a commonly accepted figure for the 'tolerance dose' in 1943, when the table was first prepared, is now well above the maximum permissible level, and due allowance must be made for this.

Activity	0·2 MeV	0·5 MeV	0·8 MeV	1·0 MeV	1·5 MeV	2·0 MeV	2·5 MeV	3·0 MeV	4·0 MeV
10 mc.	−·20	− ·71	− ·95	− ·98	− ·83	− ·61	− ·33	− ·11	+ ·19
20 mc.	−·14	− ·36	− ·27	− ·11	+ ·37	+ ·77	+ 1·15	+ 1·40	+ 1·70
50 mc.	−·07	+ ·11	+ ·63	+ 1·03	+ 1·95	+ 2·61	+ 3·10	+ 3·39	+ 3·69
100 mc.	−·01	+ ·46	+1·31	+ 1·90	+ 3·14	+ 3·99	+ 4·57	+ 4·90	+ 5·20
200 mc.	+·04	+ ·82	+1·99	+ 2·77	+ 4·34	+ 5·38	+ 6·05	+ 6·40	+ 6·70
500 mc.	+·12	+1·28	+2·89	+ 3·91	+ 5·92	+ 7·21	+ 7·99	+ 8·40	+ 8·69
1 c.	+·17	+1·64	+3·57	+ 4·78	+ 7·11	+ 8·60	+ 9·47	+ 9·90	+10·20
2 c.	+·23	+1·99	+4·25	+ 5·64	+ 8·31	+ 9·98	+10·94	+11·41	+11·71
5 c.	+·30	+2·46	+5·14	+ 6·79	+ 9·89	+11·82	+12·89	+13·40	+13·70
10 c.	+·36	+2·81	+5·82	+ 7·66	+11·08	+13·20	+14·37	+14·91	+15·21
20 c.	+·41	+3·17	+6·50	+ 8·52	+12·28	+14·59	+15·84	+16·42	+16·71
50 c.	+·49	+3·63	+7·40	+ 9·67	+13·86	+16·42	+17·79	+18·41	+18·71
100 c.	+·54	+3·99	+8·08	+10·53	+15·05	+17·81	+19·27	+19·91	+20·21
Danger range	plus	plus	plus	plus	plus	plus	plus	plus	plus
20 cm.	+·26	+1·64	+3·16	+ 4·02	+ 5·55	+ 6·44	+ 6·85	+ 7·00	+ 7·00
50 cm.	+·11	+ ·71	+1·36	+ 1·73	+ 2·39	+ 2·77	+ 2·95	+ 3·01	+ 3·01
1 m.	·00	·00	·00	·00	·00	·00	·00	·00	·00
2 m.	−·11	− ·71	−1·36	− 1·73	− 2·39	− 2·77	− 2·95	− 3·01	− 3·01
5 m.	−·26	−1·64	−3·16	− 4·02	− 5·55	− 6·44	− 6·85	− 7·00	− 7·00
10 m.	−·37	−2·35	−4·52	− 5·76	− 7·94	− 9·21	− 9·80	−10·01	−10·01

Working time	plus	plus	plus	plus	plus	plus	plus	plus	plus
1 hr./day	−·17	−1·06	−2·04	−2·60	−3·59	−4·16	−4·42	−4·52	−4·52
2 hr./day	−·11	−·71	−1·36	−1·73	−2·39	−2·77	−2·95	−3·01	−3·01
4 hr./day	−·06	−·35	−·68	−·87	−1·20	−1·39	−1·47	−1·51	−1·51
8 hr./day	·00	·00	·00	·00	·00	·00	·00	·00	·00
24 hr./day	+·09	+·56	+1·08	+1·37	+1·89	+2·20	+2·34	+2·39	+2·39

Absorber	times	times	times	times	times	times	times	times	times
Pb	1·00	1·00	1·00	1·00	1·00	1·00	1·00	1·00	1·00
Fe	8·80	2·88	1·96	1·74	1·49	1·43	1·47	1·48	1·59
Al*	41·67	9·80	6·18	5·33	4·83	5·00	5·28	5·68	6·39
H_2O	106·84	21·54	13·42	11·59	10·36	11·11	11·19	12·11	12·78

* Or concrete.

DOSE LEVELS TO THE INDIVIDUAL

(Reproduced by permission, from *The Hazards to Man of Nuclear and Allied Radiations.* Medical Research Council Report, June 1956. Cmd 9780, H.M.S.O., page 80.)

(*a*) In conditions involving persistent exposure to ionising radiations, the present standard, recommended by the International Commission on Radiological Protection, that the dose received shall not exceed 0·3 r. weekly, averaged over any period of 13 consecutive weeks, should, for the present, continue to be accepted.

(*b*) During his whole lifetime, an individual should not be allowed to accumulate more than 200 r. of '*whole body*' radiation, in addition to that received from the natural background, and this allowance should be spread over tens of years; but every endeavour should be made to keep the level of exposure as low as possible.

(*c*) An individual should not be allowed to accumulate more than 50 r. of radiation *to the gonads*, in addition to that received from the natural background, from conception to the age of 30 years; and this allowance should not apply to more than one-fiftieth of the total population of this country.

BIBLIOGRAPHY

This bibliography is not intended to be exhaustive, but merely to give a set of references to which anyone may turn who is interested in any of the topics touched upon in this book.

The following is an approximate guide to the references given:

A. Control of health hazards: nos. 1, 10 19, 26, 27, 28, 32, 34, 37, 47,

B. Laboratories for radioactive work: nos. 10, 14, 16, 24, 25, 29, 35, 39, 40, 41, 45, 46.

C. General information on the use of radioactive isotopes as tracers: nos. 1, 7, 8, 13, 17, 20, 36.

D. Radiobiology: nos. 2, 3, 5, 6, 22, 33, 44.

E. Instrumentation and physics: nos. 1, 9, 11, 12, 15, 18, 21, 23, 30, 31, 38, 42, 43.

1. *A Short Course in Radiological Protection* (1956). Health Physics Division and Isotope School. Atomic Energy Research Establishment, Harwell, Berks. Edited by R. J. Sherwood and H. J. Dunster.

2. Bacq, Z. M. and Alexander, P. (1955). *Fundamentals of Radiobiology*. London: Butterworth's Scientific Publications.

3. *Biological Hazards of Atomic Energy* (1952). Edited by A. Haddow. Clarendon Press, Oxford, being the papers read at the Conference, convened by the Institute of Biology and the Atomic Scientists Association.

4. Boursnell, J. C. and Gleeson-White, M. H. (1957). *Nature, Lond.* CLXXIX, p. 54.

5. *British Journal of Radiology* (1947). Supplement No. 1, *Certain Aspects of the Action of Radiation on Living Cells.* London: British Institute of Radiology.

6. *British Medical Bulletin* (*Radiobiology—Experimental and Applied*) (1946), **4**, no. 1. Medical Department, British Council, 3 Hanover Street, London, W. 1.

7. Calvin, M., Heidelberger, C., Reid, J. C., Tolbert, B. M. and Yankwich, P. E. (1949). *Isotopic Carbon*. New York: John Wiley and Sons. Techniques in its measurement and chemical manipulation.

8. Cook, G. B. and Duncan, J. F. (1952). *Modern Radiochemical Practice*. Oxford: Clarendon Press.

9. Cork, J. M. (1950). *Radioactivity and Nuclear Physics*. 2nd ed. New York: D. van Nostrand.

10. Dunster, H. J. (1954). The protection of personnel working with radioactive materials and the disposal of contaminated waste. *Medicine Illustrated*, **8**, 731.

11. Evans, R. D. (1948). Fundamentals of radioactivity and its instrumentation. In *Advances in Biological and Medical Physics*, **1**, 151. New York: Academic Press Inc.

12. Evans, R. D. (1947). Radioactivity units and standards. *Nucleonics*, October 1947.

13. Francis, G. E., Mulligan, W. and Wormall, A. (1954). *Isotopic Tracers*. A theoretical and practical manual for biological students and research workers. University of London Athlone Press.

14. Garden, N. B. (1949). Semihot laboratories. *Industr. Engng. Chem.* **41**, 237.

15. Glasstone, S. (1952). *Sourcebook on Atomic Energy*. London: Macmillan and Co. Ltd.

16. Henriques, F. C. and Schreiber, A. P. (1948). Administration and operation of a radiochemical laboratory. *Nucleonics*, 2, no. 3.

17. Hevesy, G. (1948). *Radioactive Indicators*. Their application in biochemistry, animal physiology and pathology. New York: Interscience Publishers Inc.

18. Hollander, J. M., Perlman, I. and Seaborg, G. T. (1953). *Table of Isotopes*. *Reviews of Modern Physics*, **25** (2), p. 469.

19. *Introductory Manual on the Control of Health Hazards from Radioactive Materials*. Issue No. 2 (January 1949). Prepared for the Medical Research Council by the Ministry of Supply, Atomic Energy Research Establishment.

20. Kamen, M. D. (1957). *Radioactive Tracers in Biology*. An introduction to tracer methodology. 3rd ed. New York: Academic Press Inc.

BIBLIOGRAPHY

21. Lapp, R. E. and Andrews, H. L. (1955). *Nuclear Radiation Physics*, 2nd ed. London: Sir Isaac Pitman and Sons, Ltd.

22. Lea, D. E. (1946). *Actions of Radiations on Living Cells*. Cambridge University Press.

23. Lenihan, J. M. A. (1954). *Atomic Energy and its Applications*. London: Sir Isaac Pitman and Sons, Ltd.

24. Levy, H. A. (1946). Some aspects of the design of radiochemical laboratories. *Chem. Engng. News*, **24**, 3168.

25. Levy, H. A. (1949). Remodelling a laboratory for radiochemical instruction or research. *Industr. Engng. Chem.* **41**, 248.

26. Morgan, K. Z. (1947). Tolerance concentrations of radioactive substances. *J. Phys. Coll. Chem.* **51**, 984.

27. Morgan, K. Z. (1947). Health physics and its control of radiation exposures at Clinton National Laboratory. *Chem. Engng. News*, **25**, 3794.

28. Morgan, K. Z. (1948). Protection against radiation hazards and maximum allowable exposure values. *J. Ind. Hyg. Toxicol.* **30**, 286.

29. Norris, W. P. (1949). Radiobiochemical Laboratories. *Industr. Engng. Chem.* **41**, 231.

30. Nuclear Data. U.S. Department of Commerce, National Bureau of Standards. Circular No. 499 (1950).

31. Parker, H. M. (1948). Health-physics, instrumentation and radiation protection. *Advances in Biological and Medical Physics*, **1**, 223. New York: Academic Press Inc.

32. *Precautions in the Use of Ionising Radiations in Industry.* Factory Dept. Ministry of Labour and National Service. Factory Form 342. H.M.S.O. (1954).

33. *Proceedings of the International Conference on the Peaceful Uses of Atomic Energy*, Geneva (1955). New York, United Nations (1956). Vol. XI, pp. 213–18. Comparative Studies of the Biological Effects of Ionising Radiation and of radiomimetic Chemical Agents.

34. *Recommendations of the International Commission on Radiological Protection* (revised 1 December 1954). *British Journal of Radiology*, Supplement no. 6. London: British Institute of Radiology (1955).

67

RADIOACTIVE TRACERS

35. Rice, C. N. (1949). Laboratory for preparation and use of radioactive organic compounds. *Industr. Engng. Chem.* **41**, 244.

36. Schweitzer, G. K. and Whitney, I. B. (1949). *Radioactive Tracer Techniques.* New York: D. van Nostrand Co. Inc.

37. Sievert, R. M. (1947). The tolerance dose and the prevention of injuries caused by ionising radiations. *Brit. J. Radiol.* **20**, 306.

38. Siri, W. E. (1949). *Isotopic Tracers and Nuclear Radiations.* New York: McGraw-Hill Book Co. Inc.

39. Solomon, A. K. and Foster, C. A. (1949). A hood for work with radioactive isotopes. *Analyt. Chem.* **21**, 304.

40. Some selected American references on the disposal of radioactive wastes. D.S.I.R. H.M. Building Research Station, Garston, Watford, Herts. Library Bibliography No. 171 (March 1957).

41. Swartout, J. A. (1949). Research with low levels of radioactivity. *Industr. Engng. Chem.* **41**, 233.

42. Taylor, D. (1951). *The Measurement of Radio Isotopes.* London: Methuen and Co. Ltd.

43. Taylor, D. and Peacock, A. G. (Editors). *Radio Isotope Instrumentation and Accessories* (1955). London: Scientific Instrument Manufacturers Association of Great Britain, Ltd.

44. *The Hazards to Man of Nuclear and Allied Radiations.* Medical Research Council. Cmd. 9780. H.M.S.O. (1956).

45. Tompkins, P. C. (1949). A radioisotope building. *Industr. Engng. Chem.* **41**, 239.

46. Tompkins, P. C. and Levy, H. A. (1949). Impact of radioactivity on chemical laboratory techniques and design. *Industr. Engng. Chem.* **41**, 228.

47. Winteringham, F. P. W. (1955). Radioactive Tracing Part X— Laboratory Discipline. *Laboratory Practice,* December, p. 493.

Abstracts of papers in this field appear bi-monthly in *Nuclear Science Abstracts.* Published by the United States Atomic Energy Commission. Technical Information Services Extension, Oak Ridge, Tennessee, U.S.A.